ETOSHA

NATIONAL PARK

DAVID ROGERS

STRUIK

CONTENTS

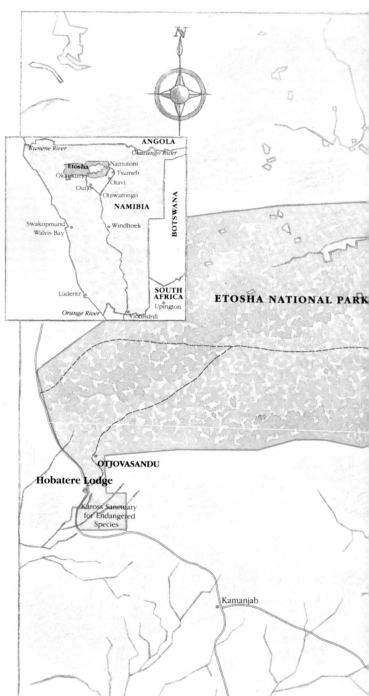

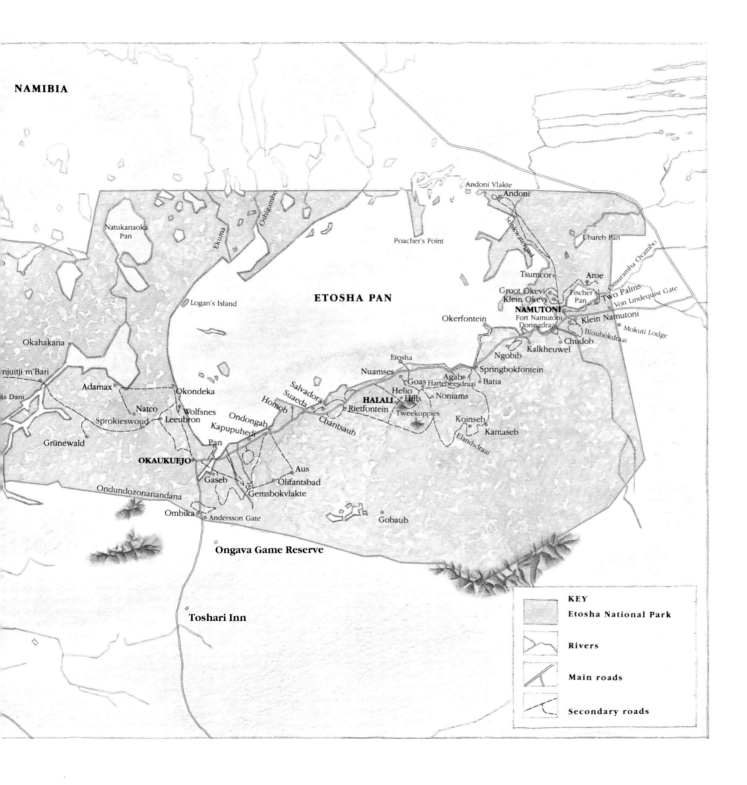

NAMIBIA

Natukanaoka Pan

Ekuma

Oshigambo

Andoni Vlakte
Andoni

Poacher's Point

Uhareb Pan

Omuramba Ovambo

ETOSHA PAN

Tsumcor

Aroe

Logan's Island

Groot Okevi
Klein Okevi

Fischer's Pan

Two Palms
Von Lindequist Gate

NAMUTONI
Fort Namutoni
Doringdraai

Klein Namutoni

Mokuti Lodge

Okerfontein

Bloubokdraai

Chudob

Okahakana

Ngobib

Kalkheuwel

njuitji m'Bari

Etosha

Springbokfontein

s Dam

Adamax

Okondeka

Nuamses

Agab

Batia

Goas Hartebeesdraai

Helio
Hills

Noniams

Natco

Wolfsnes

Salvadora

Suaeda

Homob

HALALI
Rietfontein

Tweekoppies

Sprokieswoud

Leeubron

Ondongab

Charitsaub

Koinseb

Kamaseb

Kapupuhedi

Elandsdraai

Grünewald

Pan

OKAUKUEJO

Aus

Gaseb

Olifantsbad

Ondundozonanandana

Gemsbokvlakte

Gobaub

Ombika

Andersson Gate

Ongava Game Reserve

Toshari Inn

KEY

Etosha National Park

Rivers

Main roads

Secondary roads

ETOSHA NATIONAL PARK

The great white expanse

Etosha – the place of great white spaces – is an arid land of sun and salt coloured by spectacular birds, roamed by huge herds of antelope, and ruled by magnificent elephant and large prides of lion. It is a place where mankind still has a rare opportunity to see nature as it was meant to be seen.

In the northernmost reaches of Namibia, there is an oasis of life in a seemingly desolate landscape of sand, salt and sun. It is called Etosha National Park – or 'the place of great white spaces'– an area that has earned its reputation as one of the Great Game Parks of Africa with considerable ease.

Etosha National Park offers excellent year-round game-viewing. In a region harshly governed by seasons of desiccating heat followed by a short, wet period of lushness and abundance, vast numbers of animals congregate around isolated water holes for much of the year. During the long, dry winter months you can see an overwhelming diversity of bird and animal life, including the endangered black rhino, desert-adapted gemsbok, rare black-faced impala and tiny Damara dik-dik. Visibility is also very good because the landscape is flat and sparsely vegetated.

Although Etosha is perhaps at its most exciting during the dry winter months, summer provides a different, yet equally rewarding wildlife adventure. The arid, copper-coloured landscape turns verdant green and the pans provide feeding and breeding grounds for many migratory water birds, including tens of thousands of spectacular flamingos.

The unique attraction that is Etosha is perhaps best summarised by Douglas Chadwick, who wrote in the March 1983 edition of *National Geographic*: 'It just may be possible to enter the lives of a more spectacular array of creatures with greater ease and intimacy here at Etosha than anywhere else on the globe.'

Excellent gravel roads link the park's main game-viewing areas with three well-equipped tourist camps – Namutoni, Halali, and Okaukuejo – ensuring that your stay will not only be satisfying but also very comfortable.

ABOVE: *Yellowbilled hornbill are common in Etosha and survive on a diet of insects and seeds.*
OPPOSITE: *Etosha National Park offers splendid game-viewing throughout the year. The countryside is generally open and unobstructed and the wildlife is seemingly unafraid of human presence.*

THE PRESERVATION OF A NATIONAL ASSET

For centuries before Etosha was known to western civilisation, it was home to hunter-gatherers known as Heiqum, who lived in harmony with their surroundings. During the 19th century European settlers began penetrating the region and, with their superior technology, started to adapt, not their behaviour as might have been expected, but the very landscape that they had discovered.

The history of Etosha was first recorded in the book *Lake Ngami*, written in 1851 by Charles John Andersson, a Swede. He travelled through the area with Francis Galton, an Englishman who was a cousin of the more famous Charles Darwin.

Others followed, most notably the American trader Gerald McKiernan in 1876. He eloquently praised Etosha's beauty when he said: 'All the menageries in the world turned loose would not compare with the sight I saw that day.' In the same year a group of Thirstland Trekkers under Gert Alberts, 'fleeing the ungodliness of the Pretoria Government,' made their home at the Rietfontein spring near Halali. Here they stayed with permission from the local Herero chief.

Escalating hunting and pastoral practices associated with this influx soon caused problems for the previously unspoilt landscape. In 1897, a rinderpest epidemic broke out and the threat to the livestock on farms in the vicinity necessitated the establishment of a veterinary cordon. Control posts were established at the site of present-day Namutoni and Okaukuejo.

The numbers of wild animals were also dwindling, and as a result, in 1907 Von Lindequist, Governor of German South West Africa, declared the area as a game reserve. Covering an immense 93 240 square kilometres, it was twice the size of Switzerland and easily the largest game park anywhere in the world. It had as its natural borders the Kunene River to the north and the Atlantic Ocean to the west.

Von Lindequist's new reserve was ideal for the animals which could now migrate virtually unchecked. It was less so for local people who, plagued by migrating herds and rinderpest epidemics, frequently shot at elephants and roaming antelope.

In 1970, the former Etosha Game Reserve was declared a National Park and the situation was monitored more closely, which alleviated but did not solve some of the problems. During the sixties land was reclaimed to accommodate tribal homelands, and the reserve was reduced to its present size of 22 270 square kilometres. Significantly, these new boundaries were no longer merely lines on a map, and by 1973 were solidly represented by three-metre high fences.

Following the completion of the fences, animal populations declined. The fences and atriificial water holes affected migration patterns and caused localized overgrazing. To control the latter, rotational water holes are now used. This leads to small-scale migration and ensures that areas around remaining water holes are not denuded.

Periodically conservationists are also faced with the problem of animal

Fort Namutoni

For centuries, the naturally occurring spring at Namutoni made it an important cattle post for Herero herders, who called it Omutjamatunda, or 'the place of strong waters'.

In 1897, Namutoni became a border post to control the rinderpest epidemic which was decimating domestic animals in the vicinity. Six years later the first fort was built a little way from the spring. It was to last for only one year. On 28 January 1904, the Herero Rebellion broke out and the German garrison stationed there was attacked by 500 Ovambos. Eighty of the attackers were killed in the first wave of the battle before they retreated, leaving the small garrison to contemplate its options.

The sergeant, medical orderly, two privates and three local farmers had only 150 rounds of ammunition left when they decided to flee under cover of darkness. The following day the fort was looted and razed to the ground. The Herero were defeated at the Battle of Waterberg in August 1904 and three months later a garrison of 30 men reoccupied the area. In 1905, the fort was reconstructed under Count Wilhelm von Saurma-Jetsch. A year later Namutoni became the administrative centre for the area with the count as the first commissioner.

In 1910, Namutoni was transferred to the police but two years later was again abandoned – this time for economic reasons. After the defeat of German troops in South West Africa in 1915, Namutoni was captured by South African troops.

In 1938, more trouble was in store for the small fort when lightning struck one of the towers and it was decided then to demolish the building. Fortunately, organisations and individuals contributed funds towards its preservation. In 1950 it was declared a national monument and by 1957 renovations were completed and it was opened as a tourist camp.

overpopulation. During the 1970s, the high incidence of anthrax and artificial water holes in summer-grazing areas led to a relatively high lion population. Culling of lion had not worked very well in other game parks and a better method of control was required. A small number of lionesses were treated with contraceptives by placing a time-release hormone in their necks. However, these experiments did not reduce lion numbers significantly and were discontinued in 1985.

Until recently over 65 000 black rhino roamed throughout Africa, but today they number as few as 3 000. The reason for this is economic: poachers used to sell the horns to some middle-eastern countries where they are made into knife-handles. Rhino horn is still in great demand in some eastern countries where it is used for its medicinal properties, mainly to reduce fever and in the treatment of other illnesses. There are about 300 of these endangered animals in Etosha and they are guarded very closely.

Etosha has developed a unique rhino identification system which involves monitoring all water holes on moonlit nights during the dry season. Rhino must drink daily and when they do, they are photographed, identified, and recorded in dossiers at the Etosha Ecological Institute based at Okaukuejo. Animals are also protected by the Anti-Poaching Unit, which is sometimes mounted on horseback, while some rhinos have been moved to areas in the national park where they are less likely to be poached.

Today both black and white rhino can be found in Etosha National Park. At the time of writing (February 1994), the latter have been reintroduced to the Ongava Private Game Reserve, which is located to the south of the park.

ABOVE RIGHT: *Mounted patrols of the Anti-Poaching Unit seek out armed and threatening intruders in the reserve. The unit, established in December 1988, focuses its attention on curbing commercial poaching and has been very effective in reducing the number of animals that fall victim to poachers.*

RIGHT: *This lioness maintains her dignity despite a collar placed on her to alert game rangers to her every movement.*

Anthrax – a bacterial disease affecting herbivores, particularly elephant, blue wildebeest kudu and Burchell's zebra – is another of the problems facing conservationists in Etosha. Responsible for the dwindling herds of grazers in the park, it is passed down the food chain by birds of prey such as the vulture. Lion, leopard and other predators appear to have an immunity to this disease.

Anthrax-infested carcasses used to be burnt, but this practice has been discontinued because of a shortage of wood – it takes several truckloads of wood to dispose of a dead elephant.
Nearly every natural death in the park is tested for anthrax, a threat which requires careful control, if conservationists are to maintain the all-important balance of nature in Etosha.

ETOSHA PAN - *The Living Fossil Lake*

If it was possible to have visited Etosha Pan two million years ago, you probably would have been greeted by a massive lake supporting a diverse population of birds and animals. Between two and 10 million years ago, however, the lake started to dry up at a rate of 3 000 millimetres a year. Scientists ascribe this dramatic change to the rubbing and shifting of continental plates which caused massive changes to the earth's topography. These movements shifted the course of the Kunene River, which once fed the lake, northwards and into the Atlantic Ocean. Wind erosion stripped the dry lake of its fertile loam, leaving behind a sunken depression of clay saturated with alkaline mineral salts.

Today, Etosha Pan covers 4 590 square kilometres of the 22 270 square-kilometre park. At its longest point, it is 120 kilometres across, and at its widest, it measures 72 kilometres.

Although unscientific, no theory about the formation of Etosha Pan evokes as much sentiment as that legend recounted by the indigenous people of the area, the Heiqum. They tell of a band of intruders who strayed into their territory and were surrounded by hunters who killed the men

and children, sparing only the women. One of the women sat under a tree and wept. Her tears created a huge lake which, after it dried, left a barren wasteland traced with salt: Etosha Pan.

To everyone, Etosha Pan is 'the great white expanse'. Dry for most of the year, it has not filled to capacity in recorded history. Every few years, however, heavy rains restore it to something resembling its former glory when the swollen Ekuma and Oshigambo rivers in the north-west and the Omuramba Ovambo River in the east spill into the dry pan.

Although it dominates the park and is miraculous during wet summers, Etosha National Park is not really about the pan but what happens on its fringes where nutrient-rich grazing and water is found. The pan is, however, never desolate. Animals use it as a short cut during summer migration, while for ostriches it provides a refuge at night from would-be predators.

OPPOSITE: *Swollen with summer rain, the Ekuma River spills its life-giving waters into Etosha Pan. Water adds vivid colour to the barren landscape, and is full of minerals and micro-organisms in which tiny brine shrimps and snails thrive and multiply, attracting a host of water birds which migrate to the pan to feed and to breed.*
LEFT AND BELOW: *Parched by months of drought, the pan's surface patchwork of salt-encrusted mud sustains little life, but for large herds of Burchell's zebra it is an important predator-free path to fertile grasslands.*

Although the addition of water holes and construction of fences has increased the number of animals resident in Etosha, the thin southernmost strip on the edge of Etosha Pan has been a natural oasis for animals for thousands of years during the dry winter months. This miracle can be explained by the three different types of naturally occurring water supplies.

Contact springs are found at the southern edge of the pan where water-laden limestone comes into contact with the pan's impervious clay. As the water rises, it is forced to the surface in brackish pools. Examples of these are found in Salvadora, Okerfontein and Springbokfontein.

Artesian springs – which are usually situated on top of a hillock of limestone – are created over thousands of years. Water is forced to the surface under pressure and the limestone, which is in suspension, is deposited at the surface. These are found at Klein Namutoni and Koinachas.

Water-level springs are created when subterranean water comes into contact with the earth's surface, usually as a result of a collapsed underground cave or ground subsidence. Ngobib and the two Okevi water holes are examples.

As the natural migration routes of the animals have been cut off by Etosha's fences, the naturally occurring, brackish water is no longer sufficient to support them, and it has had to be augmented by water holes fed by windmills or solar pumps.

When Animals Drink in Etosha

Although early morning and late afternoon are usually considered the best times to see game, research conducted by the Namibian Ministry of Wildlife, Conservation and Tourism reveals that different animal species in Etosha have preferred drinking times and that these extend throughout the day and night.

Lion choose the cool evening to drink, while many herbivores feel safer in the heat of the day when predators are less likely to ambush them. The following list of preferred drinking times – although not definitive – will enhance your chances of seeing the animals of your choice.

Okaukuejo's floodlit water hole provides you with a unique opportunity to watch this procession throughout the day or night. There is, however, also great merit in applying this knowledge at other water holes in the park. Spending a day at one particular water hole with a packed lunch is often more rewarding and relaxing than driving around from water hole to water hole in search of game. It is also a good idea to ask the camp attendants if you can examine their sightings book and to give you advice on which water holes they recommend you visit.

	PREFERRED	PEAK
LION	*evening*	17h00 – 23h00
ELEPHANT	*evening*	19h00 – 20h00
BLACK RHINO	*evening*	20h00 – 21h00
GIRAFFE	*all hours*	18h00 – 19h00
ZEBRA	*afternoon*	12h00 – 13h00
WILDEBEEST	*midday*	12h00 – 13h00
SPRINGBOK	*daylight*	12h00 – 13h00
GEMSBOK	*daylight*	12h00 – 13h00
KUDU	*morning*	09h00 – 10h00
HARTEBEEST	*morning*	08h00 – 16h00
ELAND	*morning*	09h00 – 10h00
WARTHOG	*afternoon*	14h00 – 15h00
OSTRICH	*daylight*	11h00 – 12h00
HYAENA	*pre-dawn*	00h00 – 01h00
JACKAL	*evening*	*sunrise , sunset*

TOP LEFT: *Etosha changes mood after the first summer rains. As the first green grass shoots appear amid carpets of yellow daisy-like* Hirpicium gazanioides *flowers, herbivores such as giraffe, zebra and springbok celebrate the start of the season of plenty.*

FAR LEFT: *After the first summer rains, bullfrogs emerge briefly from the sun-baked mud and gather in the shallow pans to breed. They remain underground for long periods if rain does not fall.*

ABOVE: *Elephant love to play in water and in mud which protect them from the searing heat. These animals need to drink daily and are dependent on the park's life-giving springs and water holes like this one at Goas, which is often frequented by elephant.*

LEFT: *While summer brings abundant water, hunting is generally easier for lion in winter, when animals must drink from water holes.*

Water is life in drought-stricken Africa and when the first summer rains begin to fall on earth that has been parched by months of sun, Etosha National Park is transformed. Dust is washed from the copper-coloured trees, green shoots appear, and water collects in pools. The desert is reborn as a verdant paradise, ripe for the arrival of birds and of babies.

The dramatic and spectacular seasonal fluctuations between the wet summer and dry winter in Etosha offer visitors two totally different, yet equally exceptional viewing opportunities.

Winter (May to August) is when most people choose to visit the park, because then the climate is pleasantly mild and the animals are most easily seen.

Temperatures are warm during the day, even in mid-July. You might even suffer from sunburn if you do not take precautions such as wearing a hat and using sun screen. The nights and early mornings are cold, however, and you can expect temperatures as low as 6 °C.

During winter, man-made water holes and natural springs are the only source of water for herds of zebra, wildebeest and impala. This focal point of activity makes it easy for both tourists and predators to find game. Lion ambush their thirsty prey from behind trees and low shrubs. Although many people believe that mornings and evenings are the best times to see game, it is well worth spending an entire day at just one water hole. You will be surprised at the procession of animals that come to drink even during the heat of midday. It is also easier for visitors to spot animals in the leafless winter vegetation than it is when they take refuge from the heat in the fullness of its summer growth.

To enjoy the splendour of Etosha during summer (November to April), you will need more heat-tolerance and lots of patience. The summer months in the park are quieter, being the less popular time for most visitors. Temperatures in excess of 35 °C are common: it is advisable that visitors keep up their intake of liquids, wear a hat and sunglasses and apply sun screen lavishly. There are frequent late-afternoon thunderstorms which bring relief from the heat and the unmistakable scent of damp dust in the air. These thunderstorms account for almost all of the precipitation in the park. Average rainfall varies from east to west over a distance of approximately 300 kilometres from 450 millimetres in Namutoni to 300 millimetres in Otjovasandu.

Because of the abundance of water in summer, many animals disperse from their traditional dry season water holes to take advantage of better grazing in the more remote northern and western regions of the park. Concentrations of game diminish which makes viewing more difficult. To offset the disappointment of not seeing as many animals as winter visitors do, it should be borne in mind that few events can compare with the wonder of watching a wildebeest dropping her young and nursing it as it takes its first few tentative steps – an event that occurs only in summer.

Equally spectacular are the migratory water birds and other waders which are attracted to the park from all over the continent after the first summer rains. These include the greater and lesser flamingo, great crested and blacknecked grebe as well as 12 species of ducks.

TOP AND OPPOSITE ABOVE: *The dry and the wet. The Goas water hole photographed in the wet summer and the dry winter shows the dramatic differences between the two seasons. Winter is traditionally the best time to view game, while summer is the time of water birds, lush vegetation and when animals drop their young.*

LEFT: *Newborn springbok lambs walk only a few hours after birth; but it's a vulnerable time for both mother and infant.*

ABOVE: *Male masked weavers are polygamous and take on dramatic plumage for the summer breeding season.*

VEGETATION – THE CLUES FOR GAME VIEWING

Etosha's landscape varies no more than 500 metres in altitude, yet its varied soil types sustain many different types of vegetation. Although interesting in itself, a study of these areas also helps you to know where game is likely to be found in the greatest numbers.

Although they are the largest geographical feature in the park, the saline pans are sparsely vegetated. Yet after the rains even these seemingly desolate areas are covered in a thick film of blue-green algae and halophytic (salt-loving) grass species such as *Sporobolus salsus*, which start to grow after the water has dried up and provide an important source of grazing for wildebeest and gemsbok during the dry winter season.

Along the edges of the pan – where water is abundant – the soil is brackish and sustains hardy halophytic plants such as the small salt bush (*Suaeda articulata*) which gave its name to the water hole near Halali where it occurs.

Further away from the pan, vegetation turns to dwarf shrub savannah which is characterised by small shrubs and the common water thorn (*Acacia nebrownii*) whose brilliant yellow flowers give colour to the park between July and September. Interspersed among the dwarf shrub is sweet grass, the single most important plant community for grazers in the park. This protein-rich vegetation thrives on loamy clay lime soil. The sweet-grass areas extend along the southern rim of the pan and it will be worth your while to spend much of your time in this area which is frequented by browsers and predators.

Three particularly complementary species are found here. The first of these is the zebra which, equipped with a hardy digestive system, tends to mow down the coarse upper stalks of grass. Next in line is the blue wildebeest which selects the lower grass blades. Its four-chambered stomach allows it to extract all available nutrients from a small amount of fodder. Finally, there is the dainty springbok which grazes on new grass shoots and, as a last resort, digs for roots.

The mopane savannah and woodland areas in the south make up the largest plant community in the park. Extending along the fringes of the pan and on the western side of the park, they comprise 80 per cent of its vegetation. It is not widely populated during winter because of the limited water supply, but it is an important grazing area in summer.

In the area around Namutoni, the tall purple-pod terminalia (*Terminalia prunioides*) and the tamboti (*Spirostachys*

africana) provide an important food source for some of Etosha's larger browsers such as giraffe and kudu, as well as the diminutive Damara dik-dik. These woodlands are interspersed with open grassy plains.

Although small in numbers, moringa trees (*Moringa ovalifolia*) are found to the west of the pan. These trees, also known as African phantom trees, are usually found on rocky hillsides, rather than on level ground. A stand of these tubular-shaped trees can be found at *Sprokieswoud* (the Enchanted Forest) near Okaukuejo.

ABOVE: *The contact spring at Salvadora. Water is forced to the surface where the water-laden limestone comes into contact with the pan's impervious clay.*
RIGHT: *This yellow flowering acacia attracts bush squirrels and a variety of browsing herbivores.*
OPPOSITE LEFT: *Parts of Etosha are surprisingly well-wooded and trees such as mopane, leadwood and tamboti abound.*

TOP: *Caked white with mud from their cooling midday wallow and satiated with up to 200 litres of water, elephant return to feed from the acacia bush south of the pan.*

ABOVE: *Standing a full six metres, giraffe are the tallest animals on earth. Pumping blood to the brain up a neck nearly three metres high, they have developed very large hearts which help maintain a blood pressure twice that of man.*

THE HERBIVORES OF ETOSHA

A wide variety of herbivorous animals have learned to survive in the arid conditions of Etosha National Park. One of the smallest of these is the tiny Damara dik-dik which weighs little more than five kilograms, while the biggest, the elephant, has an average weight of three-and-a-half tonnes. Large adults can reach a weight of six tonnes.

Gemsbok are perhaps the best-adapted of all to the semi-arid conditions. They are able to survive with little water thanks to a blood cooling system that reduces the need to perspire. Gemsbok have a network of surface blood vessels in the nose; blood is cooled by the inhaled air on its way to the brain. By allowing blood temperature in the rest of the body to increase, the need to perspire is reduced.

Elephants – which require a daily intake of approximately 200 litres of water – survive arid conditions surprisingly well. They can smell water from afar and will not hesitate in travelling great distances to slake their massive thirst. Although Etosha is now fully fenced, these man-made boundaries usually do not deter a thirsty elephant.

Today, their numbers are growing to such an extent that they are threatening the entire park's vegetation in their quest for enough food. The elephant of Etosha are curious in that their tusks are more often broken off at the base than those found elsewhere. This is probably a result of mineral deficiencies in their diet, making the ivory very brittle and is one of the reasons that these elephant have not been sought after by poachers.

Giraffe also suffer from mineral deficiencies and sometimes can be seen sucking bones, apparently aware of the need to maintain their phosphate levels. Giraffe are the tallest land animals and in this sparsely vegetated reserve they have a fairly consistent supply of food from the uppermost branches of trees which are inaccessible to other game. Although food is readily accessible to the lofty giraffe, water is less so and when stooping to drink they are extremely vulnerable to attack.

Black rhino – one of the rarest mammals in Africa – are well represented in the park and are often seen at the Okaukuejo water hole at night. They are far more aggressive than white rhino and unlike the latter have a pointed mouth which is suited for browsing. Both rhino are grey in colour and the name 'white' derives from 'wyd' which is Dutch for wide-mouthed. White rhino have recently been reintroduced into the park.

Etosha supports both Burchell's and Hartmann's mountain zebra; the former species are abundant throughout the park while the latter are found only in the western area, in the hills around Otjovasandu.

In the early morning and at midday you will notice that zebra face the sun differently in order to maintain a constant blood temperature. This adaptation to high temperatures is made possible by a variation in the ratio of white and black stripes. The rump area has a higher white to black ratio and is more reflective so the zebra points its tail to the sun when it wishes to cool down and its side to the sun to absorb heat.

FAR LEFT: *Burchell's zebra can be distinguished from the mountain variety by fewer and broader stripes which have shadows between them.*
LEFT: *Large spiralling horns are dramatic weapons of the male kudu. They are, however, seldom used by these gentle animals.*

Etosha has a large resident lion population, so if you are keen to see one of these animals you are unlikely to be disappointed. Lion certainly remain the single most popular sighting for tourists and the income they generate for the national park is essential to its survival. Lion numbers in the park vary from 200 to 350 and are ocasionally very high in relation to the numbers of antelope and zebra on which they prey. In order to understand their population dynamics, they are very carefully monitored through branding and radio tracking.

Lion are the largest of all Africa's predators – and those in Etosha are particularly big specimens. The male is most impressive – with a thick mane of black hair and weighing up to 225 kilograms, he appears much more fearsome than the female which rarely exceeds 150 kilograms and has no mane.

By virtue of her more agile frame as well as the social dictates of the pride hierarchy, the lioness is almost exclusively responsible for hunting. This, however, does not stop the male lion from being first in line to eat when the prey has been brought down. The male has an important role, however, in ensuring that the pride is not threatened by other predators.

Lion are the most sociable of the cats and the breeding prides can have as many as 30 members, though prides of this size are not often encountered. Bachelor groups generally tend to be much smaller. The cubs are reared by their parents for the first two years, but starvation among the young accounts for a high mortality rate of up to 60 per cent in Etosha.

Lion are opportunistic animals and although they generally hunt at night, they will also attack during the heat of the day if they have a chance of doing so without expending too much energy.

You are most likely to see lion at water holes such as Okaukuejo, Okondeka, Goas, Rietfontein, Charitsaub, Gemsbokvlakte and Ombika. At night you can hear their roar up to 9 kilometres away – a chilling reminder to other predators and prey just who is 'king of the beasts'.

Leopard are solitary and elusive cats, making them a rare and exciting sight for visitors. Because they are nocturnal, they are less likely to be seen than the other large cats, which hunt during the day.

They have magnificent rosette-shaped markings – a pattern which has been emulated by the manufacturers of camouflage uniforms around the world – which makes them very difficult to spot.

Leopard are extremely athletic and often carry their weighty prey into trees out of reach of other predators such as hyaena. They are easily able to carry small antelope such as impala, but there are records of small giraffe weighing up to 90 kilograms being transported by leopard. The lofty larder serves as a useful hiding place for the solitary hunter when it sets off in search of water. Their diet is extremely varied and ranges from squirrels to antelopes.

Leopard are a little more than a third of the size of lion, with males weighing up to 90 kilograms and females up to 60 kilograms. Because of their elusive nature, an accurate population census is difficult to achieve, but park officials estimate their numbers at over 100. The young are reared for two years, but join in the hunt – albeit more as a spectator than as a participant – when they are as young as two months old.

The area around Namutoni is the best place to see leopard, particulary the two Okevi water holes, Klein Namutoni and Kalkheuwel. Although less prevalent, they also occur towards Halali and Okaukuejo near the Rietfontein and Goas water holes.

The smallest – but perhaps most spectacular – of all the large cats in Etosha is the cheetah. It is the fastest of all land animals and can reach speeds of up to 110 kilometres per hour, although it is unlikely to sustain this for more than about 300 metres.

At full charge, the gracefully built cheetah appears to be the perfect running machine. With its head held dead still and its legs pumping like pistons, it uses its tail to act as a counter balance as it follows the turns and swerves of its fleeing prey.

Although it has similar markings to the leopard, the cheetah's spots are completely round. It also has a far smaller body and its weight seldom exceeds 60 kilograms. A distinctive feature is the large tear markings which run scimitar-like beneath each eye. It is quite likely that these cut down glare in much the same way as the paint used by baseball players helps them catch balls against the full sun.

There only seem to be between 50 and 100 cheetah in Etosha. Competition from other predators such as lion and hyaena who often chase the smaller cheetah from their prey appears to keep their numbers down. This problem is compounded because the number of grazers in the park is proportionally low as a result of the killer-disease, anthrax, a long spell of poor rainfall and other factors.

Despite Etosha's relative scarcity of cheetah, they are fairly easy to spot as they tend to favour open, savannah grasslands. The best places to see them are on the plains near Leeubron-Adamax water hole, Halali and on the Andoni plains.

OPPOSITE: *The white-washed, saline deposits near Suaeda provide a playground for cheetah and a startlingly beautiful background for photographers.*
ABOVE: *Although leopard are usually found in the low-lying branches of high trees surveying their domain, Etosha is not very well treed, so sightings are more likely to be at ground level, in the shade, or on rocky outcrops.*
RIGHT: *Lion spend up to 20 hours a day sleeping or dozing in the shade of trees or in long grass. With the onset of evening, they become more active and this is when they are most likely to be seen drinking at water holes.*

Etosha is a bird-lover's paradise with over 340 species having been identified by ornithologists. Although some of these are found here all year round, others migrate from other parts of Africa, and Europe to feed, sometimes to breed and are only found here during the summer. Whether migratory or permanent residents, many of these birds have adapted fascinating features for survival of their species.

Ostrich (*Struthio camelus*), the world's largest bird, are a common sight on the edge of the pan, grazing on the nutrient-rich grasses in close proximity to the water holes. This flightless bird is a match for any predator, as it can reach a speed of up to 60 kilometres per hour and has a clawed toe which can seriously injure a predator. During breeding season, the ostrich lays a clutch of eggs, each of which is the size of 24 chicken eggs. This is followed by six week incubation.

A bird that occurs in the dry thornveld surroundings is the crimsonbreasted boubou (*Laniarius atrococcineus*). It has striking red and black markings which are reminiscent of the old German flag which once flew over Namutoni. Equally impressive is the lilacbreasted roller (*Coracias caudata*) which is found in the park all year round.

The pygmy falcon (*Polihierax semitorquatus*) measures only 150 millimetres from tail to beak – the smallest of all raptors – and one of 35 species found in the park. One of the peculiarities of pygmy falcons is their propensity to occupy nests of the sociable weaver (*Philetairus socius*).

Of all the desert-adapted birds, perhaps the most interesting are the Namaqua sandgrouse (*Pterocles namaqua*). These seed-eaters are often forced to make a daily trip of up to 60 kilometres from their nests to the nearest water. The flock, which can number 300 birds, departs at precisely timed intervals after sunrise when the temperature is cool and they will expend least energy.

After the first summer rains, huge flocks of greater and lesser flamingo (*Phoenicopterus ruber* and *minor*) congregate on the pans.

OPPOSITE ABOVE: *The bulky bateleur eagle's distinctive rocking motion in flight gives it its name, which is derived from French and means a tightrope walker.*
OPPOSITE BELOW LEFT: *The brightly coloured crimsonbreasted boubou is a striking resident of the arid thornveld.*
OPPOSITE BELOW RIGHT: *Namaqua sandgrouse wade into water in order to trap moisture in their chest feathers, where it is stored for transportation back to their young.*
BELOW LEFT: *In flight, the secretarybird resembles an eagle or vulture at first glance, but it can be identified by its very long legs which protrude well beyond its elongated central tail.*
BELOW RIGHT: *The doublebanded courser blends in well with Etosha's sombre winter hues. These birds generally leave the area when it takes on its verdant green summer complexion.*

NAMUTONI

The place of strong waters

With a dusty, burnt-red sun against a woodland silhouette, the haunting

call of the last post reverberates through the African bush, signalling the end of

another day. This is the sunset flag lowering ceremony at Namutoni and a visit

to Etosha is not complete without having witnessed it.

Namutoni – 'the place of strong waters' – is possibly the most scenic of the three camps in Etosha. Surrounded by fan palms, it is centred around a Beau Geste-style fort, reminiscent of the days of German colonialism. Situated on a small rise, it has an elevated tower from which you can survey the wild tamboti and terminalia woodland peculiar to this area of the park. This is leopard country and also home to rare animals such as the black-faced impala and the diminutive Damara dik-dik.

In the late afternoon if you take the Dik-dik drive (also called 'Bloubok Drive') – an 11-kilometre round trip from Namutoni – you may be fortunate enough to see both of these antelope in the sheltered tamboti woodland. Only the male dik-dik has horns and both sexes weigh less than 5 kilograms. They are very skittish and when startled will retreat in short, stiff-legged, backward 'stots' which are accompanied by a series of short, high-pitched whistles. Together, this sounds not unlike 'dik-dik' and may account for their name. A characteristic of the dik-dik is the dark-coloured gland area which is situated under the base of each eye. The dik-dik rubs this gland against plants, and secretes a hormone to mark its territory.

Fischer's Pan – named after Lieutenant Adolf Fischer who was appointed the first game warden of the park in 1907 – is well worth a visit. Situated immediately north of Namutoni, it is reached by way of a circular drive. The pan is particularly dramatic in summer when flamingo, pelican and other water birds are abundant. Along the route to Fischer's Pan you will pass Two Palms, one of the most spectacular water holes in the park. Only 11 kilometres from Namutoni, it is ideally located for sunset photography, but you should make sure that you leave sufficient time to get back to the camp before the gates are locked. Another water hole that provides outstanding game

OPPOSITE: *A lone giraffe, ever aware of the danger of predators, takes a tentative drink from Klein Namutoni water hole.*
ABOVE: *Damara dik-dik are most commonly seen in the Namutoni region of the park where the dense woodland provides both food and shelter.*

viewing and photographic opportunities is that situated at Chudop. Like Namutoni and Klein Namutoni, this is a natural water hole fed by an artesian spring. The water is forced up through the limestone and emerges on top of an outcrop. Its floating reed-island is characteristic of this type of spring. Chudop is frequented by giraffe and is one of the few places where you are likely to see black-faced impala, and eland.

Namutoni is the second-oldest camp in Etosha and, in addition to the allure of the fort, it has a museum which is worth a visit.

The arrival of flamingos in Etosha is a spectacle witnessed by few and forgotten by none. When the November rains fill Etosha and Fischer's Pan, they are transformed into a kaleidoscope of pink and white as tens of thousands of these splendid birds arrive to feed and to breed. They fly from all over Africa and can cover hundreds of kilometres in a day.

Flamingos are cautious by nature, and they will only set about their breeding rituals if there is sufficient water for them to rear their young before the pans dry out. Their nests are constructed from mud and stand about 30 centimetres tall with a hollow crown which holds a single egg. After an incubation period of 28 days the baby flamingos, or pulli, start to hatch in their thousands. The pulli are grouped together in communal crèches for protection and, by some miracle of nature, parents are able to single out their own offspring from the multitude by the sound of their call. The chicks are fed by their parents for the first 65 to 75 days until they are able to fly and fend for themselves. This is a critical period for the flamingos of Etosha and often they are caught in a race against time as the waters of the pan begin to evaporate with the onset of summer.

In 1971, 30 000 pulli undertook a 70-kilometre trek from a dry Etosha Pan to the

Ekuma Delta, while their parents flew constant relays of food to sustain them on their arduous journey. It was a sight that is still recalled by the management at Etosha, who assisted by trucking large numbers of the birds to safety. Operation Flamingo was a success and 25 000 of the birds that started the trek survived.

Both greater and lesser flamingos occur in Etosha and intermingle freely. Proportionately, they have the longest neck and legs of any birds. These features assist them to stand upright while swinging their heads about under the water, filtering small particles of food – an action not unlike an underwater vacuum cleaner. Their beaks are also perfectly suited for the task. Bent and flattened towards the end, the beak's fine comb-like teeth (or lamellae) remain parallel to the ground while the bird's head is in the upside-down feeding position.

Although both types of flamingo are similar in appearance, they have different feeding patterns. The greater flamingo stamps its way through the shallow waters and filters out small aquatic invertebrates from the disturbed mud. The lesser flamingo, on the other hand, walks or swims while feeding on blue and green algae and phytoplankton which are suspended in the water.

OPPOSITE ABOVE: *Namutoni is a well-established camp with a swimming pool, restaurant, shop and other modern facilities to enjoy after a long and rewarding day's game viewing.*

ABOVE: *Thousands of flamingos migrate to Fischer's Pan near Namutoni following the heavy summer rains. Etosha is one of the only areas in Africa where these birds breed in large numbers.*

LEFT: *Etosha National Park is one of the few places where the rare black-faced impala may be seen. This antelope favours savannah woodland, avoiding open grassland unless there is bush cover.*

ABOVE: *The black rhino, one of the most threatened large mammals in Africa, is a bad-tempered animal that will charge with very little provocation. Huge middens of dung mark the borders of its territory, which it patrols regularly for unwelcome intruders.*

RIGHT: *A blue wildebeest strikes a lonesome sight as it makes its way across Etosha Pan. These animals are usually found in large herds which afford them greater protection against predators.*

OPPOSITE TOP: *The black-backed jackal is a stealthy animal that remains easily undetected as it approaches the water's edge for its early morning drink. It lives on a varied diet of mice, birds, insects, berries and seeds.*

OPPOSITE BELOW: *South of Namutoni rest camp is the popular Dik-dik drive where, in the late afternoon, you have a good chance of seeing these tiny antelope with their distinctive and protruding mobile muzzles.*

HALALI

The call of the bugle

With some of the sweetest grazing to be found along the perimeter of Etosha Pan, the Halali area is home to large herds of zebra, wildebeest and a variety of other antelope, which attract a host of attendant predators.

Established in 1967, Halali is the newest, smallest and also the quietest of the camps in Etosha. Situated midway between the camps at Okaukuejo and Namutoni, it is built at the foot of the Tsumasa Hill near Twee Koppies or Helio Hills, the highest point in the flat eastern and central areas of Etosha.

Halali takes its name from the German expression for the traditional bugle call sounded at the end of a hunt. Helio Hills, on the other hand, harks back to the days when a German heliograph unit was stationed there early this century.

There is plenty to see and do in and around Halali. The vegetation includes mopane trees and sweet-grass veld growing on a lime soil. The latter provides protein-rich grazing for many herbivores. Herds of Burchell's zebra, blue wildebeest, springbok and other antelope are common, and where you find these animals, you can be sure that predators such as lion, cheetah and leopard are not far behind.

The water hole at Goas is well worth a visit. It is a very attractive game-viewing area and within easy striking distance of the camp. Here you are likely to see raptors such as the martial and tawny eagles, and the bateleur. Among the animals which may be encountered are elephant, Burchell's zebra, lion, blue wildebeest, red hartebeest, and, if you are fortunate, the black-faced impala.

The dark reddish-brown, black-faced impala is very similar in appearance to its more common relative, but is slightly heavier and can be distinguished by the black flash which runs down the centre of its nose and beneath each eye. Northern Namibia is the southernmost area of their limited distribution which extends only as far as Angola. Less than 1 000 black-faced impala inhabit Etosha.

All the camps have a sightings book and before setting out on your journey, it is worth perusing to see where you are most

OPPOSITE: *A late afternoon thunderstorm brings relief from the summer heat and the scent of damp on the Halali plains.*
ABOVE: *Although they are generally grazers, springbok also relish browsing on tasty acacia blossoms.*

likely to find concentrations of game. The area along the pan to the west of Halali is noted for its cheetah and lion sightings. The stark-white saline deposits at Suaeda and Salvadora pans provide a wonderful washed-out background for photographing these animals.

The camp at Halali is a restful, shady place with green lawns, a swimming pool, shops and a restaurant. A nature walk has been laid out through the dolomite hills inside the camp confines, and after a long drive, you can stretch your legs and enjoy the birdlife of the mopane woodland.

Halali is probably the best place in southern Africa to see barecheeked babbler.

These birds are quite tame here and can be seen in large groups in the camp and the surrounding rocky wooded hills, emitting low, continuous grating 'chuck chuck' sounds. The barecheeked babbler is about 24 centimetres tall and has a white body with a black tail and a brown band on its neck that runs down the length of its back.

An attractive viewpoint on the camp's perimeter overlooks a water hole which, like the one at Okaukuejo, is floodlit at night. It offers rewarding game-viewing during the winter months.In close proximity to the Rietfontein water hole west of Halali is a tumble-down ruin which is said to have been the home of the Thirstland Trekker

leader, Gert Alberts. Alberts and his party reached the area in January 1876 having survived months of hardship while travelling through Bechuanaland (Botswana). Although he maintained that the purpose of his journey was because 'trekking was in our hearts', he was deeply concerned that the British would annexe the Transvaal and about 'the ungodliness of the Pretoria government under T.F. Burgers, as well as the freeing of coloured people, and other unacceptable laws'. Alberts' wife died here and there is a memorial near the old house, which was eventually abandoned when the trekkers moved on to establish themselves in present-day Angola.

LEFT: *Leopards are usually solitary cats, seen in pairs only when they come together to mate or when a female has cubs. Although generally active at night, they can sometimes be spotted during the cooler daylight hours.*

ABOVE: *Lion normally hunt larger mammals, particularly ungulates, but they will attack anything from mice to young elephant. They also scavenge and will chase other predators, especially cheetah, from their kills.*

TOP LEFT: *White-throated or rock monitors are adept at climbing, having sturdy legs and strong claws, and are at home in trees or on top of rocky outcrops. They have a long tongue which is used as a sensory device, and feed on small mammals, insects, birds and eggs. Their natural enemies include large raptors.*

TOP RIGHT: *The nocturnal scrub hare - the largest and longest eared of all hares on the southern African continent. It inhabits woodland and scrub areas.*

OKAUKUEJO

The handle of the well

The spectacular water hole at Okaukuejo is, for many visitors, the most lasting

impression of Etosha National Park. After all, where else on earth

can you view such an interesting procession

of wildlife 24 hours a day?

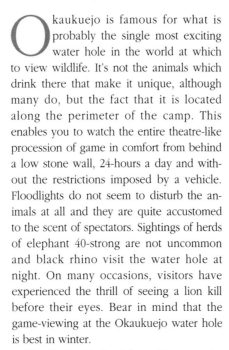

Okaukuejo is famous for what is probably the single most exciting water hole in the world at which to view wildlife. It's not the animals which drink there that make it unique, although many do, but the fact that it is located along the perimeter of the camp. This enables you to watch the entire theatre-like procession of game in comfort from behind a low stone wall, 24-hours a day and without the restrictions imposed by a vehicle. Floodlights do not seem to disturb the animals at all and they are quite accustomed to the scent of spectators. Sightings of herds of elephant 40-strong are not uncommon and black rhino visit the water hole at night. On many occasions, visitors have experienced the thrill of seeing a lion kill before their eyes. Bear in mind that the game-viewing at the Okaukuejo water hole is best in winter.

Okaukuejo (derived from Okoquea, the name for the handle on a water-well's pulley) is the largest of the three camps in Etosha and is also the headquarters of the Etosha Ecological Institute which is responsible for the management and control of

vegetation and animal numbers. It also runs the only facility for full-time anthrax research in Africa.

The camp has excellent accommodation in bungalows, amenities for caravanners and campers, and boasts a post office, museum, two swimming pools, shops and a restaurant. If you are interested in identifying Etosha's animal spoor, there is a fascinating display of footprint impressions in a concrete slab just outside the Etosha Ecological Institute (which is closed to the public). You can also climb up the tall tower above the camp complex for a wonderful view of the park and the distant Ondundozonananandana Mountains which form part of the southernmost border of the park. The name means 'the place where young calves go and do not return', a roundabout way of explaining to young

OPPOSITE: *The water hole at Okaukuejo is one of the best in Etosha for general game-viewing, with elephant one of its regular visitors.*

ABOVE: *A pair of secretarybirds stand vigil on their bush-top nest.*

cattle and sheep herders that they should beware of resident leopards. Like Namutoni, Okaukuejo was originally established as a control post after the outbreak of the rinderpest epidemic in 1897. It, too, once had a fortified watchtower, constructed in 1901 out of limestone, which served as a police post. This was destroyed a few years later and was never rebuilt. The first game warden at Etosha, Lieutenant Adolf Fischer was stationed in the camel stables behind the present-day restaurant and his first guests pitched their tents around a water hole. In 1957, Okaukuejo became Etosha's first tourist camp.

The vegetation immediately around the camp is mainly protein-rich sweet limeveld but as you drive north-west towards Okondeka, this changes into the shrub mopaneveld and finally to the Okondeka duneveld. This is lion country; the Okondeka water hole provides the only shade for approximatelyt 5 kilometres and the lion frequently show their authority by taking up position under these sought-after trees. You are also likely to see many other animals, including giraffe and kudu, at the water hole, which is well suited for afternoon photography.

Some 37 kilometres distant from Okaukuejo lies *Sprokieswoud* – or the Enchanted Forest – which is populated by 900 *Moringa ovalifolia* trees. The forest is unique because it is the only area where these plants occur on a flat landscape.

Although this has long mystified botanists, Dr Hu Berry in his book *Etosha National Park*, offers the following explanation of the phenomenon. According to Heiqum legend, 'The Great God in his act of creation had found a place on earth for all animals and plants. His task completed, he realised that he still held a number of moringas in his hand and, not knowing where to plant them, flung them, roots pointing skyward, into their present location.'

Giraffe feed from the upper leaves of the moringa and, to protect them from these herbivores as well as from elephant, a section of the forest has been closed off and the water supply to nearby artificially-fed water holes in the area has been cut off.

OPPOSITE ABOVE: *Animals visiting Okaukuejo's spectacular water hole are accustomed to people viewing them from behind the low stone wall.*

OPPOSITE MIDDLE: *Large, very sensitive ears enable the bat-eared fox to detect insects burrowing far beneath the soil surface. Neither true fox nor jackal, the bat-eared fox, which stands a mere 30 centimetres tall, is the only creature to belong to the genus Otocyon .*

OPPOSITE BELOW: *The sociable, yellow mongoose lives in elaborate underground warrens with up to seven members and several entrances. These gregarious animals are often found in association with ground squirrels and sometimes share their warrens.*

ABOVE: *Okaukuejo's luxurious setting and facilities offer a most comfortable and relaxing post-bush experience.*

LEFT: *The Moringa Forest or* Sprokies-woud *– a stand of 900* Moringa ovalifolia *trees – lies to the west of Okaukuejo and is well worth a visit. Porcupines are known to favour the bark of these trees.*

TOP: *Dark blue clouds gathering on summer afternoons promise rain and relief from the searing heat.*

ABOVE: *The well-developed olfactory senses of the short-snouted elephant-shrew offer its prey little hope of escape. This solitary, mainly diurnal animal lives on a diet of insects and other invertebrates, with a preference for ants and termites. Its name is derived from its elongated, trunk-like snout.*

RIGHT: *Lion are opportunistic animals and although they generally hunt at night, they will also do so during the heat of the day as long as they do not have to expend too much energy.*

OPPOSITE TOP: *This unfortunate giraffe was killed by a pride of lion. Having eaten the better part of the animal, black-backed jackal have been permitted to take their share of the kill.*

THE WESTERN AREAS

The place of young men

A single gravel road and a sanctuary for endangered species are the only

developments in the wild, western reaches of Etosha National Park.

For the adventurous traveller, it offers the opportunity to

experience a unique wilderness region.

Otjovasandu – 'the place of young men' – is the gateway to the western area of Etosha and, as its name suggests, is not for the faint-hearted. Few tourists are permitted to venture into this undeveloped section of Etosha, although there are plans to establish a fourth camp in the area when sufficient funds become available. Only authorized tour operators are allowed to enter the park from the west.

The drive through this virginal landscape follows the dust road that traces, and has been dubbed, the 19th latitude. The way is long and very bumpy, and there are no signs of civilisation for over 100 kilometres until you arrive at Okaukuejo.

Those who venture into the west will be justly rewarded. The topography and vegetation is quite unlike that of the rest of the park. The savannah grassland and terminalia and acacia woodlands are scattered with numerous granite boulders and dolomite hills, and there are mountains in the vicinity of Otjovasandu. While the topography and vegetation is different, so too are the animals. In this area, you may see the rare Hartmann's mountain zebra

alongside their more widely distributed cousins, Burchell's zebra. The former number 900 and are restricted to the mountainous areas surrounding Otjovasandu.

Another rare animal to be found in this area is the eland. About 500 occur naturally in the park and they mostly favour the remote north western areas. Weighing up to 850 kilograms, they are the heaviest and largest of all antelope species and one of the most dramatic to behold. Also occurring here is the park's only troop of baboon, comprising less than 50 members.

Unlike the area to the east, which is fed by natural springs that run along the southern edge of the pan, the west has little non-porous clay to force water to the

LEFT: *Granite boulders, more resistant to wind and water erosion than the surrounding landscape, are typical features of the western areas near the Kaross rare species camp.*
ABOVE: *Nowhere in Etosha are cheetah numerous; they are generally found on the open plains, but can also be seen in the western area of the park.*

surface and there are few naturally occuring water holes. Few artificially fed water holes have been constructed and, as a result, the area is sparsely populated during winter. This all changes with the onset of the summer rains, when zebra and blue wildebeest migrate from the denuded eastern grasslands into the verdant pastures to the west. They leave a trail across the pan which is followed by other animals, including a host of predators.

Although the edge of the reserve is guarded by a fence, constructed from sturdy railway lines in places, elephant migrate north quite freely into Kaokoland, Damaraland and Kavango to return in winter when water is scarce. In the wet season, there are up to 1 200 elephant in the park, but this number increases to some 2 000 during the dry months. Elephants in Etosha are in no short supply; the park is periodically over-

populated by these magnificent, yet destructive animals, but anthrax usually limits their numbers. Ironically, between the turn of the century and the 1950s, elephant were entirely absent from the park and only returned when artificially constructed water holes were created.

In an effort to conserve and breed the rare animals of Etosha National Park, a 15 000-hectare enclave called the Kaross rare species camp was founded during the 1970s in the extreme south-west of the national park. The camp serves as a holding and breeding facility for endangered and less common species such as the black rhino, black-faced impala, roan antelope and eland. From here the animals are being relocated to other suitable areas of Etosha and Namibia. The camp is not open to the general public. More like a farm than a game reserve, Kaross has a number of

artificially-fed water troughs for the animals and the area is fenced off for safety, free from predators and also the competition from common species.

ABOVE: *Gemsbok are ideally adapted to the arid western areas. Although they will drink readily when water is available, they are able to survive for long periods without drinking.*

OPPOSITE ABOVE RIGHT: *Roan, second largest antelope after the eland, are not endemic to Etosha National Park; but they are bred at Kaross rare species camp from where they are relocated to game parks in the eastern Caprivi Strip.*

OPPOSITE RIGHT: *During summer when water and vegetation is plentiful, huge herds of zebra migrate to the western areas which are less well-populated by predators and visitors.*

GUIDE TO WATER HOLES IN ETOSHA

The concentration of animals at Etosha is spectacular, and the dependence on water holes throughout much of the year makes the wildlife readily visible. The following is a list of the main water holes together with a brief description of the animals you are likely to see:

Adamax: A seasonal water hole situated north of Okaukuejo this water hole has been dry in recent years.

Andoni: Etosha's northernmost water hole lies in the Andoni plains. It's more like a vlei and is very good for birdwatching.

Aroe: Situated on the northern edge of Fischer's Pan, this is an excellent site for viewing flamingo and other water birds in summer. In winter it is likely to attract elephant, springbok, blue wildebeest, kudu, zebra and giraffe.

Aus: This brackish water hole is fed by an artesian spring and is likely to provide sightings of elephant, zebra, springbok, red hartebeest, kudu and gemsbok. If you are fortunate, you may see black rhino and black-faced impala as well. It is well worth a visit.

Batia: This water hole is recommended for general game viewing. Elephant, springbok, blue wildebeest and gemsbok are likely to be seen.

Charitsaub: Lion are known to frequent this very open area. You are also likely to see springbok, zebra and gemsbok in large numbers.

Chudop: Eland are scarce in the park, but you may see them mid-morning at this artesian spring which is characterised by its floating reed island. The water hole is famed for its giraffe and provides excellent photographic opportunities.

Gemsbokvlakte: The water hole is excellent for game viewing in winter, including lion and elephant. It is a good site for afternoon photography.

Goas: One of the best water holes for general viewing in Etosha during the winter months, it is frequented by raptors and a vast number of animals, including lion, elephant, zebra, red hartebeest and black-faced impala.

Great Okevi: Cheetah, leopard, elephant, kudu and zebra are some of the larger animals found at this water-level spring.

Helio: This small water hole provides an opportunity to see elephant and other game at close quarters.

Homob: Located roughly halfway between Halali and Okaukuejo and far from the road, this water hole is not ideal for photographs, but lion and elephant, may be clearly viewed with the aid of binoculars.

Kalkheuwel: This is a photographer's paradise. You can get very close to a wide variety of game from two sides of the water hole. Lion, gemsbok, giraffe, zebra, springbok, elephants, leopard, and raptors, including the bateleur, tawny eagle, and black kite, may be seen.

Koinachas: Although it does not always provide game-viewing opportunities, this water hole is frequented by a wide variety of animals.

Kapupuhedi: When there is water here, you will have an outstanding view of game against the stark white background of the pan. It is frequented by gemsbok, zebra, impala and wildebeest.

Klein Namutoni: Most animals – including elephant, giraffe, zebra, leopard and the black-faced impala – may be seen at this natural spring on the road to Dik-dik drive. In the late afternoon, keep a look out for the tiny Damara dik-dik in the vicinity. This is also the best time for photography at this water hole as the light is excellent.

Klein Okevi: This is an attractive water hole which is used by a great variety of birds and animals, including black-faced impala, kudu, gemsbok, elephant and cheetah.

Leeubron: Mostly dry, this seasonal water hole has been known to attract huge zebra herds and large numbers of lion. It is an excellent spot for photography when it contains water.

Namutoni: This artesian spring is surrounded by dense reeds and does not attract much game.

Ngobib: Although this water-level spring is in a sunken depression and does not lend itself well to game viewing, it does attract leopard, kudu, zebra and elephant which may be seen in the vicinity.

Noniams: Although not as impressive as nearby Goas, this water hole attracts several species of game.

Nuamses: A wide variety of game, including elephant, may be seen at this water hole on the edge of the pan.

Okaukuejo: This famous water hole is situated within the camp and is flood-lit at night, giving visitors a unique opportunity to view game at all hours of the day and night. Elephant, black rhino, lion, giraffe and other game frequent the water hole during winter.

Okondeka: Famous for its pride of lions and also frequented by a host of other animals, including kudu and giraffe, this water hole is well worth a visit and offers excellent opportunities for afternoon photography.

Okerfontein: Although it has the distinction of being the strongest contact spring in the park, this water hole is not particularly well-frequented by wildlife.

Olifantsbad: As its name suggests, this water hole is popular with elephant herds. Zebra, many species of antelope, including black-faced impala and red hartebeest, and warthog may be found here.

Ombika: Although far from the road, this is a well-used drinking place frequented by a large variety of animals and birds.

Ondongab: This natural water fountain is often dry, but when it does have water, animals, including giraffe, may be seen against the white backdrop of the pan.

Ozonjuitji m'Bari: This water hole is fed artificially from a borehole, providing water to a variety of game including black rhino and elephant. The rare roan antelope may also be seen here.

Rietfontein: One of the largest water holes in the park, this is a must for bird-watchers. Nearly all the species in Etosha may be seen here; it is also a good position for general wildlife photography.

Salvadora: Lion, cheetah, springbok, zebra and gemsbok are found at this busy water hole on the edge of the pan.

Suaeda: Situated alongside Salvadora, it provides similar game viewing against a white saline background.

Springbokfontein: This natural contact spring at the edge of the pan is not very popular with game.

Tsumcor: If you want to take photographs of elephant, then a visit to this water hole is a must. A wide variety of game can be found here.

Two Palms: Situated on the edge of Fischer's Pan, this water hole is dominated by its two makalani palms and provides excellent opportunities for photographs at sunset. Make sure you leave enough time to travel the 11 kilometres back to camp before the gates are locked.

ETOSHA AT A GLANCE

Etosha has three public rest camps – Namutoni, Halali and Okaukuejo – but if you wish for more personalised service, you may consider staying in one of the privately run camps and hotels as your springboard into the park. It is essential to book accommodation well in advance, even if you only plan to camp or caravan in the park.

GETTING THERE

Namutoni is reached by following the B1 through Otjiwarongo, Otavi and Tsumeb, then proceeding towards Etosha's Von Lindequist gate, 553 kilometres from Windhoek. The camp is 11 kilometres further on. Okaukuejo can be reached via the B1 to Otjiwarongo, then turning onto the C38 to Outjo which will take you to the Andersson gate, 447 kilometres from Windhoek. Halali is accessible via Namutoni and Okaukuejo and is 75 kilometres and 70 kilometres from them respectively. Ensure that you arrive before sunset when the gates are still open.

Namutoni, Halali and Okaukuejo each have an airstrip; private pilots must make prior arrangements with the camp authorities before landing there. At Namutoni, a private tour operator will arrange transport to and from the airstrip only for the guests of Etosha Fly-In Safaris. Pilots should circle the camps to announce their arrival. There is no aviation fuel available in the park, but you can make arrangements to obtain 200-litre drums from Mokuti Lodge.

Only organised tour operators are permitted to enter the park from the west or travel from Okaukuejo to Otjovasandu.

RESERVATIONS

Applications should be addressed to the Director of Tourism, Reservations, Private Bag 13267, Windhoek, Namibia. Telephone (061) 3-6975 for reservations; (061) 3-3875 for information. Facsimile for both is (061) 3-3845. Applications should include type of accommodation required, dates of arrival and departure and the number of adults, and number and ages of children.

ACCOMMODATION AND FACILITIES

Okaukuejo, Halali and Namutoni offer two-, three- and four-bed self-catering bungalows which are equipped with fridges, kettles and washbasins. They also have hot-plates, and showers and toilets except for the two-bedded one room accommodation at Halali, where guests use communal kitchen and ablution facilities. No cutlery or crockery is provided. Halali also has four-bed accommodation in tents. Namutoni offers two-, three- and four-bedded rooms in the historic fort and also has four-bed mobile homes. All three camps have a camp or caravan sites (each site accommodates a maximum of 8 people and 2 vehicles), plus special accommodation for bus groups and picnic sites for day-visitors. Each camp has a restaurant (serving meals from 07h00 – 09h00, 12h00 — 14h00, and 18h00 – 20h30) as well as a snack kiosk and a shop which sells provisions, ice, souvenirs and books. They also each have a filling station, swimming pool, mailing facilities, and there is firewood available.

Hobatere Lodge

Hobatere is run by dedicated conservationists and is ideally suited to discerning game enthusiasts and wildlife photographers. The lodge is situated in a 35 000-hectare reserve near Otjovasandu, the gateway to the undeveloped western portion of Etosha National Park. There are 11 *en suite* units, each accommodating 2 people, and facilities include a central dining area, boma and swimming pool. There is also a rustic camp within the reserve that can accommodate a further 16 people.

Casual visitors are not permitted entry to the Otjovasandu area from Hobatere, but some tour operators use this lodge. To get there from Windhoek, take the B1 to Otjiwarongo and then the C38 to Outjo. About 9 kilometres beyond Outjo turn left along the C40 to Kamanjab. Continue through the town, taking the C35 towards Ruacana. The Hobatere signpost is 65 kilometres beyond Kamanjab. From there the lodge is a further 15 kilometres away. The road is tarred until Kamanjab. Hobatere has its own airstrip.

For more information, contact Hobatere Lodge, PO Box 110, Kamanjab, Namibia. Telephone (06552) 2022.

BELOW LEFT: *Halali rest camp, equidistant to both Okaukuejo and Namutoni, is noted for its many large mopane trees, which provide welcome shade for visitors.*
BELOW: *Well-equipped rondavels provide comfortable accommodation at Okaukuejo rest camp.*

Mokuti Lodge

Mokuti Lodge is a luxury hotel situated a short 12-km drive from the camp at Namutoni. From here you can either drive your own car into the park or join one of the organised tours laid on by Namib Wilderness Safaris which is based at the lodge. Mokuti boasts air-conditioned chalets each with its own telephone and can accommodate 216 guests in total. The grounds are landscaped with shady trees, lawns and a swimming pool. Other facilities include two restaurants and a boma in which you can enjoy traditional African fare, two short walking trails, a 2-kilometre trim track, a visit to the reptile park and guided horse trails. The lodge is set in 100 hectares of natural bush and game has been introduced to its newly established Klein Begin Game Reserve.

The lodge will arrange transport from Tsumeb airport where scheduled flights arrive from Windhoek's Eros airport. There are flights in a 19-seater Beechcraft turbo-prop aircraft from Eros directly to the airstrip at the lodge on Tuesdays and Thursdays. Mokuti is reached by road from Windhoek by following the B1 through Otjiwarongo, Otavi and Tsumeb and then proceeding towards Etosha. The lodge is situated 500 metres before the Von Lindequist gate which is the eastern entrance near to Namutoni. The road is tarred all the way from Windhoek.

For more information, contact Mokuti Lodge, PO Box 403, Tsumeb, Namibia. Telephone and facsimile (0671) 2 1084.

Ongava Game Reserve

If you want to see white rhino, then Ongava is the place to try to see them. Ongava – the largest private, fenced game reserve in Namibia – recently bought six of these animals. The reserve is situated on the southern boundary of Etosha, and covers an area of 32 000 hectares at the foot of the Ondundozonananandana Mountains to the south of Okaukuejo.

Ongava is an upmarket establishment that allows you to choose the type of accommodation and safari holiday that best suits your needs. You can stay either in luxury chalets or a tented camp, or opt for the remoteness of a wilderness camp.

The luxury lodge has 10 thatched chalets with air-conditioning, *en suite* bathrooms and balconies. There is a swimming pool and reference and video libraries to enjoy. The tented camp has five, two-bedded tents while the wilderness camp comprises five, two-bedded Himba-style huts. Activities at Ongava include guided walking trails, day and night nature drives in the game reserve and into the national park. Horse trails can also be arranged.

Ongava Game Reserve has an airstrip for private or charter flights which may be arranged through your travel agent or by contacting the reserve. By road follow the B1 to Otjiwarongo, then the C38 towards Okaukuejo. Turn left off the tarmac road at the signpost to Ongava, 5 kilometres before the Ombika/Andersson gate entrance to Etosha National Park.

For more information, contact Ongava Game Reserve, PO Box 186, Outjo, Namibia. Telephone (06542) 3413.

Toshari Inn

Toshari Inn is a useful springboard into Etosha National Park. It has 16 *en suite* bedrooms which are air-conditioned and comfortably furnished, and set in scenic dolomotic hills studded with white seringas (*Kirkia acuminata*). Guests can dine in the restaurant or opt for a self-catering unit.

By road take the B1 to Otjiwarongo, then the C38 via Outjo towards Okaukuejo. Turn right off the tarmac road at the signpost to Toshari Inn, some 26 kilometres before the Ombika/Anderssen gate entrance to the national park.

For more information, write to Toshari Inn, PO Box 164, Outjo, Namibia. Telephone (06542) 3602.

ABOVE TOP LEFT: *Mokuti Lodge offers hotel luxury just outside the Namutoni entrance to the park.*
ABOVE RIGHT: *Hobatere Lodge has its own reserve and visitors can enjoy guided tours in four-wheel vehicles.*
ABOVE LEFT: *The swimming pool is one of Mokuti's more popular facilities.*

WHEN TO GO

Most people prefer to visit Etosha during the winter months (April to September). The days are warm to hot, but evenings are cooler with temperatures sometimes dropping to below freezing. At this time of year water is scarce, forcing animals to congregate in large numbers at water holes. Following the onset of the summer rains, however, is also very rewarding with lush green vegetation and the promise of seeing water birds in their thousands. For birding, Etosha is best visited during March and April when the migrants are still present. The camps are open all year round.

CURRENCY AND ENTRY REQUIREMENTS

Only cash in South African Rand and Namibia Dollar, Visa, MasterCard, Diners Club and travellers' cheques in South African Rand are accepted. Foreigners are required to be in possession of a valid passport and visa where necessary. South Africans should take note that Identity Books are no longer acceptable in Namibia as travelling documents and that their petro-cards are invalid.

GETTING AROUND

A comprehensive map of the park and all its water holes is available from all the rest camps. The roads within Etosha are gravel but in good condition and well-signposted, making driving easy for first-time visitors in vehicles with a high clearance. Vehicles which are high off the ground and have large windows are ideal for viewing game. There is an Avis office in Tsumeb with an agency at Mokuti Lodge and numerous car-hire companies operate from Windhoek.

Instead of driving yourself through the park, there are several tour operators who conduct organised expeditions to Etosha. Tours are ideal for people who have not visited African game parks before as the guides have considerable knowledge about the fauna and flora of the park.

Details of car-hire companies and tour operators are listed in the Namibia Accommodation Guide for Tourists which is published annually and is available free of charge from any Namibian Tourism Offices. For more information, write to Namibia Tourism, Private Bag 13346, Windhoek, Namibia. Telephone (061) 28-49111. Facsimile (061) 22-1930. There are also offices in Johannesburg and Cape Town in South Africa, Frankfurt in Germany and London in the United Kingdom.

WHAT TO TAKE

The clothes you take will depend on the time of year you visit the park. The average daily maximum during December is about 35 °C and the average daily minimum in July about 6 °C. In this sub-tropical climate, days during winter are quite warm, so as well as a jersey and jacket you will need light clothing, sun screen, lip salve and a broad-brimmed hat. Sunglasses are useful to reduce the sharp glare. Also bring a torch and a bathing costume.

Binoculars are invaluable game-viewing aids and you should choose any model in the 7x35 or 8x40 range, although bird-watchers may prefer the increased magnification of the 10x40 models. When selecting binoculars, the light factor should be as high as possible to ensure they can be used in low-light conditions. To calculate this divide the two numbers (eg. a 10x40 model will have a light factor of 4 and an 8x40 model a light factor of 5).

Photographic equipment should include the most powerful telephoto lens that you can afford, as game photographed with a standard lens can often be very disappointing. Longer lenses must be held steady and a tripod will be useful for the water hole at Okaukuejo, while photography from you car will be assisted by steadying your camera on a beanbag slung over a partially rolled-down window. For low-light conditions an ISO rating of 200 will be ideal and for bright light, ISO 50. Many people select the ISO 100 as a good all-round film. Ensure that your camera and film are kept in a cool place at all times.

IMPORTANT INFORMATION

Visitors to the Etosha National Park must reach camps before sunset and may only depart after sunrise. The following are prohibited: open vehicles, motor-cycles, air-guns, unsealed firearms, disturbing of animals, leaving the road and leaving the vehicle outside the rest camp, except at demarcated picnic sites. No pets are allowed. The speed limit is 60 kilometres per hour.

MEDICAL INFORMATION

Northern Namibia is a malaria area and it is particularly prevalent during summer. Consult your doctor or pharmacist before entering the park so that a suitable course of anti-malaria tablets can be followed. To avoid being bitten by mosquitos, additional precautions should also be followed, in particular, wear trousers and a long-sleeved shirt at night and apply insect repellent before sundown. Malaria rarely occurs in Etosha however.

The Namibian Tourism Department assures visitors that all blood in Namibia is donated by carefully selected unpaid volunteers only and the Blood Transfusion Service of Namibia screens all blood and blood products for various transmittable diseases including hepatitis and AIDS.

SUGGESTED FURTHER READING

Birds of Etosha National Park. R.A.C. Jensen and C. F. Clining, Directorate of Nature Conservation and Recreation Resorts, 1983
Etosha. Daryl and Sharna Balfour, Struik Publishers, 1992
Etosha National Park. Hu Berry, Struik Publishers, 1993
Guide to Namibian Game Parks. Willie and Sandra Olivier, Longman Namibia, 1993
Guide to Southern African Safari Lodges. Peter Joyce, Struik Pulbishers, 1993
The Living Deserts of Southern Africa. Barry Lovegrove, Fernwood Press, 1993
Reader's Digest Illustrated Guide to Game Parks and Nature Reserves of Southern Africa. Reader's Digest Association of South Africa (Cape Town), 1991
Sasol Birds of Southern Africa. Ian Sinclair, Phil Hockey and Warwick Tarboton, Struik Publishers, 1993
Southern African Birds: A Photographic Guide. Ian Sinclair, Struik Publishers, 1993
Southern African Snakes and Other Reptiles: A Photographic Guide. Bill Branch, Struik Publishers, 1993
Southern African Trees: A Photographic Guide. Piet van Wyk, Struik 1993
Southern, Central and East African Mammals: A Photographic Guide. Chris and Tilde Stuart, Struik Publishers, 1992

Maps
The Globetrotter's Travel Map of Namibia, Struik Publishers
Namibia Minimap, Map Studio
Namibia Traveller's Map, Macmillan
Road Map of Etosha National Park, Namibia Tourism (available at rest camps)

INDEX

Struik Publishers (Pty) Ltd
*(a member of The Struik Publishing
Group (Pty) Ltd)
Cornelis Struik House, 80 McKenzie Street
Cape Town 8001*

Reg. No.: 54/00965/07

First published in 1994

*Text © **David Rogers**
Map © Loretta Chegwidden
Photographs © individual photographers and/or
their agents as follows:*

***Daryl Balfour:** pp. 1, 2, 3, 4, 6, 9 (above and
below), 11 (above and below), 13 (above and
below), 14, 15 (above and below left and right),
16, 18 (above), 20, 22 (below left and right), 23
(right), 24, 25, 28 (above), 30 [ABPL], 31, 33
(above left and right), 35, 36 (below top
and bottom), 40, 41, 42, 43 ,
back cover. **Gerald Cubitt:** pp. cover,
10, 17 (below), 26 (below), 27,
36 (above), 39 (above), 46 (above left).
Clem Haagner [ABPL]: p. 12 (above).
Robert C Nunnington: p. 37 (above).
Willie and Sandra Olivier: p. 45 (right). ©
**Struik Image Library: Photographer Peter
Pickford:** pp. 7, 8, 12 (below), 17 (above), 18
(below left and right), 19, 21 (above and below),
22 (above), 23 (left), 28 (below), 29 (above and
below), 32, 33 (below), 38 (below), 39 (below).
Gavin Thompson [ABPL]:
p. 26 (above).
Mark van Aardt: pp. 8, 34, 37 (below), 38
(above), 45 (left46 (below left and top right).*

*Project Co-ordinator: Marje Hemp
Editorial assistants: Christine Didcott and
Simon Atkinson
Designed by René Greeff
Design assistant: Lyndall Hamilton
Typeset by Suzanne Fortescue, Struik DTP,
Cape Town
Reproduction by Hirt & Carter (Pty) Ltd,
Cape Town
Printed and bound by Kyodo Printing Co (Pte)
Ltd, Singapore*

ISBN 1 86825 606 5

Acknowledgements
*The publishers and the author wish to thank
Daryl Balfour for his invaluable contribution to
the text and photography, Willie and Sandra
Olivier for their useful comments on recent developments in the park, and Dr. Malan Lindeque of
the Etosha Ecological Institute at Okaukuejo for
ensuring that the information in this guide is
factual and up to date.*